YOUR KNOWLEDGE HAS VALUE

- We will publish your bachelor's and master's thesis, essays and papers

- Your own eBook and book - sold worldwide in all relevant shops

- Earn money with each sale

Upload your text at www.GRIN.com and publish for free

Bibliographic information published by the German National Library:

The German National Library lists this publication in the National Bibliography; detailed bibliographic data are available on the Internet at http://dnb.dnb.de .

Imprint:

Copyright © 2018 GRIN Verlag
Print and binding: Books on Demand GmbH, Norderstedt Germany
ISBN: 9783668877788

This book at GRIN:

https://www.grin.com/document/446948

Regasa Wake

The Determinants of High Yielding Wheat Varieties Adoption by Small-Holder Farmers in Ethiopia

Wheat and its Importance for the Ethiopian Food Security

GRIN Verlag

DETERMINANTS OF HIGH YIELDING WHEAT VARIETIES ADOPTION BY SMALL-HOLDER FARMERS IN ETHIOPIA

ABSTRACT

Adoption of high yielding wheat varieties is one of the measures presumed to enhance wheat yield in Ethiopia. However, there are several socioeconomic and institutional factors that limit the adoption of high yielding wheat varieties. The main objective of this study was to identify factors affecting adoption of high yielding wheat varieties in Mao-Komo district of Benishangul-Gumuz, Ethiopia. The study used cross-sectional data collected from sample of 174 farm households selected through two-stage stratified random sampling techniques. Descriptive statistics and econometric models were used to analyze the data. Probit model was employed for adoption analyze of high yielding wheat varieties. The probit model result depicted that land holding size, tropical livestock unit, access to agricultural information, frequency of extension contacts, off-farm income, and perception of farmers toward attributes of high yielding wheat varieties affected the likelihood of adoption of high yielding wheat varieties positively and significantly. But, sex of household heads, and affiliation to organizations had negative and significant effect on the likelihood of adoption of high yielding wheat varieties. The findings suggest that the government and stakeholders should need to focus on improving farm land and livestock productivity, strengthening frequency of extension visits, encouraging participation in off-farm activities, creating reliable information and awareness towards farmers' perceptions in the area. Finally, further support of high yielding wheat varieties adoption should be given due attention for smallholders.

Key words: Adoption, High yielding wheat varieties, Smallholder, Binary probit

1. INTRODUCTION

The economic development of Ethiopia is highly dependent on the performance of its agricultural sector since it is the main economic pillar of economic growth of the country. Agriculture contributes 42 % of the GDP of the country and about 85 % of the population gains their livelihood directly or indirectly from agricultural production (CSA, 2015).

Wheat (*Triticum aestivum L.*) is one of the major food and cash crops for smallholders in Ethiopia. It is important cereal crop with annual production of about 4.23 million tons and cultivated on an area of 1.66 million hectares (CSA, 2015). According to the CSA report, it occupies about 24.02 % of the total cereal area in the country and contribute the grain production about 15.65 %. However, its national average yield is about 25.43 quintals per hectare. This is low yield compared to global average of 40 quintals per hectare (FAO, 2009). The low yield has made Ethiopia unable to meet the high demand and the country is net importer of wheat (Rashid, 2010).

Wheat is one of the major cereals of choice in Ethiopia, dominating food habits and dietary practices, and is known to be a major source of energy and protein in the country (Hailu, 2003).The utilization of wheat has increased due to the growing of urbanization and the expansion of agro-industries used as raw material, and also considered to attain food security in Ethiopia. It is also used for traditional foods and the straw is used for animal feed and thatching of roofs (Katherine, 2013).

To feed the rapidly growing population and meet the high demand of wheat in the country, it needs to increase the production and yield of wheat. However, increasing yield requires successful adoption of improved agricultural technologies (Dorosh and Rashid, 2013). Low yield due to low adoption of improved agricultural technologies is believed to be the main factor that prevented agricultural production from coping with the rapid population growth in Ethiopia. For this reason technological change is commonly considered as one of the major options leading to successful productivity growth in agriculture. The objective of this study was to assess the adoption of high yielding wheat varieties by smallholder farmers.

2. Empirical Studies on Adoption of Agricultural Technologies

A number of empirical studies have been conducted by different people and institutions on farmers' adoption behavior both outside and inside Ethiopia using econometric models. The results of various empirical studies confirmed that adoption of a new technology offers opportunities for increasing productivity and production.

Assefa and Gezahegn (2010), Solomon *et al.* (2011) and Hassen *et al.* (2012) found that age of household head, educational status, livestock holding, non-farm income, sex, and information access plays important factors in affecting the decision of farmers to adopt improved technology. Mahadi *et al.* (2012) studied factors affecting adoption of improved sorghum varieties in Somali Region of Ethiopia. They have found out that more educated farmers are more likely to adopt improved sorghum varieties in the study area.

Degye (2013) conducted study on agricultural technology adoption, diversification and commercialization for enhancing food security in Eastern and Central Ethiopia by using multivariate probit revealed that adoption of high yielding crop variety was influenced by land allocated, agricultural income, distance to research institution, and the farming system. It also was reported significantly and negatively affected by other exogenous shocks. Degnet and Mekibib (2013) found that membership to farmer cooperatives has a strong positive effect on adoption of chemical fertilizer.

The study by Bekele *et al.* (2014) on adoption of improved wheat varieties and impact on household food security in Ethiopia, indicated that wheat technology adoption has generated a significant positive impact on food security and these results provide strong evidence for the positive impact of adoption of modern agricultural technologies for a major food staple on alleviating food insecurity in rural Ethiopia.

Leake and Adam (2015) conducted study on factors influencing allocation of land for improved wheat variety by smallholder farmers in Adwa district. They pointed out adopters had high family labor, high number of tropical livestock unit, large land size, high frequency of extension contact, access to credit, access to education, access to nearest to main road and market as compared to non-adopters. They also indicated that education level of household head, family size, tropical livestock, distance from main road and nearest market, access to credit service, extension contact and perception of household toward cost of the technology have to be significantly affecting factors adoption of improved wheat variety.

The study conducted by Sisay (2016) on agricultural technology adoption, crop diversification and efficiency of maize-dominated farming system in Jimma Zone of South-Western Ethiopia by using Tobit model indicated that age, family size, level of education, family education, ownership of mobile phone, extension services, cooperative membership, livestock holding and land holding size have positively and significantly influenced probability of improved maize variety and/or chemical fertilizer adoption in maize farming while, distance of development center from residence has a significant negative effect.

3. MATERIALS AND METHODS

3.1. Description of Study Area

The study was conducted in Mao-Komo Special district of Benishangul-Gumuz Region located in the Western part of Ethiopia and stretches along the Sudanese border found around 661 km away from Addis Ababa. Mao-Komo Special district is one of the 20 districts found in Benishangul-Gumuz Region, its capital, Tongo, located 112 km away from Assosa town, the capital city of the region. It is bordered by Oromia Regional state in the East, Sudan in the West, Assosa Zone in the North and Gambela Region in the South. The altitude of the district ranges from 950-1960 m.a.s.l. The temperature of the district ranges from 17.5-32 °C. The rainfall is uni-modal which starts in the month of April and ends in mid-October. The annual rainfall ranges from 900-1800 mm with mean annual rainfall is 1316 mm, mostly received between May and September with the highest in July and August. The duration is about 6 to 7 months with good amount of rainfall distribution.

Having an area of about 2100 Km^2 and population of about 42,050 (CSA, 2007), the district has 7848 households with 7185 and 663 male and female headed households, respectively. The district is mainly characterized by two agro-ecologies; namely, *"Kolla"* and *"Woina Dega"* that has been structured with 32 Kebeles that comprise 24 and 8 kebeles, respectively. From these, 5 Kebeles are the most wheat producers in the area. Farming is the predominant occupation of the people in the area since it is the main economic stay of the district. Maize, sorghum, wheat, and finger millet are the dominant cereal crops produced for consumption. Coffee, sesame, nigger seed and *teff* are produced for income generation in the district. Cattle, small ruminant, donkey, poultry and honey bee are the most important livestock species. The district has potential and favorable environmental and socio-economic conditions that would suitable to wheat production.

3.2. Methods of Data Collection and Sampling Procedures

The data for this study were collected from both primary and secondary sources on a wide variety of variables. The primary data were collected through individual interviews of selected respondents and the survey was administered using semi-structured questionnaires within individual interview. To complement the primary data, secondary data were obtained from different unpublished and archival sources such as articles/literatures, official reports, CSA report data, and personal communications.

This study defines the survey population at two levels, namely at the rural kebele level and at the farm household level. A two-stage stratified random sampling method was employed to draw representative sample respondents to increase homogeneity within adoption stratum and heterogeneity between strata. In the first stage, rural kebele administrations were stratified into two categories as potential and less potential wheat growers. Accordingly, three potential wheat producing kebeles were randomly selected.

In the second stage, members of each kebele were stratified into two groups based on their adoption status of high yielding wheat varieties. Then, a total of 174 farmers were randomly sampled (87 from each group) taking into account probability proportional to size of households in each kebele for both groups. Finally, the survey was administered and data were collected and analyzed on 174 respondents.

3.4. Methods of Data Analysis

In the study of adoption of high yielding wheat varieties through descriptive and econometric methods of data analysis were used to assess the relationship between explanatory and dependent variable. Adoption of high yielding wheat varieties was evaluated by statistical tools and econometric models for concluding the socio-economic, institutional, and environmental factors that hinder the adoption by smallholder farmers in the study area.

Descriptive statistics

Descriptive statistics were utilized to assess the socio-economic characteristics of the sample respondents for adoption of high yielding wheat varieties in the study area. These information was considered to augment the econometric analysis results. The descriptive analysis tools such as t-test and chi-square were employed to assess the relationship among the variables of interest that statistically to compare users and non-users respondents of high yielding wheat varieties.

Econometric models

In this study, the econometric analysis that employed was binary probit model for adoption of high yielding wheat varieties.

Probability of Adoption

A household level adoption study considers the decision made by the household head to include new or improved variety in usual farming practice. The decision made to adopt or otherwise depend on different factors. Farmers' decision to adopt high yielding wheat varieties is assumed to be the product of a complex preference comparison made by a farm household. To adopt or not to adopt high yielding wheat varieties is often a discrete choice. Discrete choice models have widely been used in estimating models that involve discrete economic decision-making processes (Guerrem and Moon, 2004).

The dependent variable which is normally used with these models is dichotomous in nature, taking the values 1 or 0, a qualitative variable which is incorporated into the regression model as dummy variable. In this case the value 1 indicates a farmer who adopts the high yielding wheat varieties while the value 0 indicates the farmer who does not adopt. A number of studies conducted on technology adoption of smallholder farmers indicate that for such type, the most commonly used models are the logit and probit models. The binary model, a logistic distribution function, and the probit model, a normal distribution function, is used in estimating the probability of technology adoption (Pindyck and Rubinfeld, 1981; Feder *et al.*, 1985; CIMMYT, 1993; Greene and Zhang, 2003). Such models have been widely used in different adoption studies not only to help in assessing the effects of various factors that influence the adoption of a given technology, but also to provide the predicted probabilities of adoption (Asfaw *et al.*, 1997). The estimating model that emerges from normal CDF is popularly known as the Probit model (Gujarati, 1995).

Therefore, the study utilized the probit model to analyze likelihood of adoption of smallholder farmers because it is an appropriate econometric model for the binary dependent variable and the error term is assumed to be normally distributed. Often, probit model is imperative when an individual is to choose one from two alternative choices, in this case, either to adopt or not to adopt high yielding wheat varieties. An individual makes a decision to adopt high yielding varieties of wheat if the utility associated with that adoption choice is higher than the utility associated with decision not to adopt. Hence, in this model there is a latent or unobservable variable that takes all the values in $(-\infty, +\infty)$.

In this study the classical probit model was employed to analyze the adoption decision behaviour of the respondents. To motivate the Probit model it can be assumed that the decision of a household (in this case an adopter) to adopt the high yielding wheat varieties (Yi=1) or not (Yi=0) depends on an unobservable utility index (also known as a latent variable), that is determined by one or more explanatory variables, Xi, in such a way that the larger the value of the latent variable, the greater the probability of a household to adopt the high yielding wheat varieties.

The Probit model is specified as:

Where $Y_i = X_i \beta_i + \varepsilon_i$ where i= 1, 2, 3…..n   (1)

Where: Y_i is a dummy variable indicating the probability of adoption and related as:

$Y_i = 1$ if $Y_i > 0$, otherwise $Y_i = 0$

X_i - is household characteristics of variables that determining farmers adoption in the probit model

β_i - is unknown parameter to be estimated in the probit regression model

4. RESULTS AND DISCUSSION

The study presents the descriptive results explaining smallholder farmers' probability of adoption of high yielding wheat varieties by smallholder farmers through the statistical analysis of descriptive tools and empirical results of econometric analysis.

4.1. Descriptive Results

Descriptive statistics such as mean, minimum and maximum values, range and standard deviations were used to describe the socio-economic and institutional characteristics of the households under considered in the study of high yielding wheat varieties adoption. For this study, the data was collected from both adopters and non-adopters of high yielding wheat varieties that consists of 50 % each of two group. Table 1 & 2 below, depicts the statistical t/χ^2-test comparison of variables expected to determine adoption of high yielding wheat varieties of sample households.

Table 1. Descriptive Statistics for Some selected Continues Variables

Descriptions	Adoption Status			t-test value
	Adopters	Non-adopters	Total Sample	
	Mean	Mean	Mean	
Land holding size (ha)	1.83	1.16	1.50	-5.114***
Total livestock unit (tlu)	4.29	1.46	2.88	-6.803***
Educational level (years)	1.99	1.49	1.74	-1.517*
Distance from market center (km)	4.20	4.37	4.29	0.447
Distance to main road (km)	2.83	2.77	2.80	-0.145
Farming experiences (years)	25.39	23.48	24.44	-1.165
Family size (number)	7.59	6.72	7.16	-2.100**

| Off/non-farm income (ETB) | 3.26 | 3.25 | 3.26 | -0.024 |
| Frequency of extension contact (days) | 20.48 | 12.97 | 16.72 | -3.263*** |

*, **, and *** indicates significant at 10 %, 5 % and 1 % significance levels, respectively.

Source: Own survey (2015)

The descriptive results revealed that adopters of high yielding wheat varieties were significantly different from non-adopters in many cases such as farm land holding size, family size, livestock ownership, frequency of extension visit, educational level, and perceptions' of farmers toward high yielding wheat varieties on certain attributes. On the other hand, adopters did not make significant difference in terms of distance from market center, distance to main road, farming experiences, access to credit services, sex of household head, off/non-farm income activities, and participation in local level organization with compared to non-adopters (Table 1).

Table 2. Descriptive Statistics for Some selected Discrete/ Dummy Variables

Descriptions	Adoption Status			χ^2 value
	Adopters Mean	Non-adopters Mean	Total Sample Mean	
Farmers perceptions' for HYWV				
Good	45.98	19.54	65.52	55.441***
Average	2.30	26.44	28.74	
Poor	1.72	4.02	5.75	
Access to credit services				
Yes	3.45	2.87	6.45	0.755
No	46.55	47.13	93.68	
Access to information				
Yes	43.68	31.61	75.29	13.622***
No	18.39	6.32	24.71	
Affiliation to organizations				
Yes	48.28	47.13	95.40	0.469
No	1.72	2.87	4.60	
Sex of household				
Male	47.13	47.70	94.83	0.732
Female	2.87	2.30	5.17	

*, **, and *** indicates significant at 10 %, 5 % and 1 % significance levels, respectively.

Source: Own survey (2015)

Mostly, peoples living in same environment share a common understanding of various circumstances and from more or less similar perception about certain situation. However, the degree of perception varies from individual to individual due to different factors. Adopters have more experienced advantages of high yielding wheat varieties with attributes like higher yield per hectare, short duration to maturity, better market price, and resistance to pests and diseases. The chi-square test indicated that there is systematic relationship between perception of respondents about high yielding wheat varieties and the two categories of sample household groups that adopters and non-adopters were statistically significant (Table 2).

Households' access and arrangement to institutions play a vital role in providing agricultural services like access to market information, microfinance/credit services, and information about new wheat varieties, consultancy services provided by development agents, NGOs, researchers, cooperative and other social services. There is no significant difference in terms of household average distance from nearest market center, distance to main road between adopters and non-adopters (Table 2).

4.2.   Econometric Results

Analysis of adoption

The probit model was employed to identify factors influencing adoption of high yielding wheat varieties by smallholder farmers in the study area. The chi-square ($\chi2$) distribution was used as the measure of overall significance of a model in probit model estimation. The model had a log pseudo likelihood of (-84.10) after fifth iteration. The Wald chi^2 test statistics with 13 degree of freedom is equal to 51.05, and prob > chi^2 = 0.0000 is used to test the dependence of the adoption of high yielding wheat varieties on the selected independent variables in the model (the hypothesis that all coefficients are equal to zero is rejected at 1% significance level). The pseudo R^2 (0.30) which indicates 30 % of the variation between adopters and non-adopters of high yielding wheat varieties which explained by the variables.

The results indicate that the independent variables are related to the propensity of adoption of high yielding wheat varieties at 1 % significance level, indicating that the model have good explanatory power. Hence, the adoption decision of high yielding wheat varieties by households is best explained by the probit model because the assumption of normality of the errors is supported by the gof test and statistically significant. The results of the model show that out of the thirteen variables included in the model, eight are correlated with probability of high yielding wheat varieties adoption and found to have statistically significant effects on the adoption of high yielding wheat varieties on the sample respondents. The binary probit model outputs showed that sex of household, difference in land holding size, tropical livestock unit, frequency of extension contacts, access to informations, off-farm income activities, perception of farmers' toward attributes of high yielding wheat varieties, affiliation to local organizations are significant factors affecting the probability of adoption of high yielding wheat varieties.

The probit model results revealed that household head sex is negatively and significantly associated with adoption of high yielding wheat varieties. The result confirms that as compared to male-headed households, female-headed households are less likely to adopt high yielding wheat varieties than male-headed farmers. Implication female-headed households on likelihood of adoption of high yielding wheat varieties might be that female headed households have a lower labor endowment, lower farm land holding and livestock unit ownership, and less access to information on high yielding wheat varieties compared to their counterpart.

From marginal effects, being female-headed households, *citrus paribus*, reduce by 23.6 % the adoption of high yielding wheat varieties as compared to male-headed households. In the study area, letting females to be a household head is not yet well developed and recognized since almost majority of the farmers are Muslim followers, and female-headed households mostly are those who are widowed and divorced. In such instances, due to the cultural and socio-economic factors, their likelihood of adopting high yielding wheat varieties becomes negligible. The overall finding is consistent with the results reported by others. Ouma *et al.* (2002) and Tura *et al.* (2010) pointed out a negative association between a female-headed household and improved maize variety adoption. Contrary to this, results by Menale *et al.* (2012), Solomon *et al.*(2014) and Tesfaye *et al.* (2014) found significance influence of sex of household head on adoption of technologies.

It is worth to note that, having more farm land size is one best option whereby smallholders could be prompted in diversifying their crop production and adopt all imperative yield increasing technologies. Farm size have a positive and significant effect on adoption of high yielding wheat varieties. The positive effects of farm size indicate that farmers with relatively large farm land size decide to adopt high yielding wheat varieties than owners of small farms land size. As a basic production factor, the more farmers have cultivable land, the more likely to adopt agricultural technologies particularly high yielding wheat varieties that could possibly increase crop yield. Probably, owning more arable land could be taken as a prerequisite to adopt and employ agricultural technologies since farmers incur a cost. The marginal effects indicated that as land holding of the households' increases by a unit (1 hectare), farmers' likely to participate in adoption of high yielding wheat varieties increases by 16.4 % as compared to non-adopters. During focus group discussion farmers told that shortage of farm land especially in some villages' due to expansion of urban and degradation of farm land influence the cultivation of high yielding wheat varieties. Besides this, increasing number of family members per households in the area enforce farmers to share their farm land to children that reduce the yet low land size and made difference in adoption of high yielding wheat varieties. To overcome this problem, some farmers went to *Kolla* area far away from their home residence for cultivation of low land crops for home consumption purpose. The result is supported by findings of earlier studies on technology adoption of Hailu (2008), Ahmed (2010), Yishak and Punjabi (2011) and Abreham (2012). Other studies by Jon (2007), Assefa and Gezahegn (2010), Moti *et al.* (2013) and Sisay (2016) also found a similar result.

Table 3. Estimated results of probit model likelihood of adoption

Adoption of high yielding wheat variety	Coefficient	Robust Std.Err	Marginal effects
Sex of household head (1=male; 0=female)	-0.633*	0.376	-0.236
Farming experience (years)	0.011	0.012	0.005
Educational level (years)	0.085	0.061	0.034
Distance to main road (km)	0.054	0.053	0.021
Family size (numbers)	-0.017	0.045	-0.007
Land holding size (ha)	0.413**	0.195	0.164
Tropical livestock unit (tlu)	0.193***	0.058	0.077
Access to credit (dummy)	-0.350	0.484	-0.138
Frequency of extension contact (days)	0.022**	0.009	0.009
Access to information (0=no; 1=yes)	0.971***	0.321	0.366
Off/non-farm income (ln ETB)	0.063**	0.029	0.025
Perception of farmers' toward HYWV	1.438***	0.362	0.455

Affiliation to organizations (0=no; 1=yes)	-0.768*	0.446	-0.278
Constant	-2.884***	0.822	
Number of observation	174		
Wald chi^2 (13)	51.05***		
Prob > chi^2	0.0000		
Pseudo R^2	0.3027		
Log pseudo likelihood	-84.10		

*, ** and *** indicates significant at 10 %, 5%, and 1 % significance levels, respectively

Source: Own survey (2015)

Tropical livestock unit, which is a proxy for measuring wealth status of household head (in terms of tropical livestock unit), is found to have a positive and significant influence on adoption of high yielding wheat varieties, indicating that farmers with large number of livestock are more likely to adopt high yielding wheat varieties than others. Results of analysis of marginal effect show that a unit increase in tropical livestock unit, increases the decision of high yielding wheat varieties adoption by 7.7 % of adopters of sample households. This is because farmers with relatively more livestock unit make use of their income obtained from sale of livestock and their byproducts for the purchase of high yielding wheat varieties seed for cultivation of grain wheat. Also livestock, particularly oxen, are used for draft for different farm operations. The study is supported by Hailu (2008), Solomon *et al.* (2011), Hassen *et al.* (2012), Berihun *et al.* (2014), Tolesa (2014), Leake and Adam (2015) and Sisay (2016) which confirms the same result.

The number of extension contacts per year (season) has positively and significantly related to adoption of high yielding wheat varieties, implying that farmers with more frequency of extension contacts are more likely to adopt high yielding wheat varieties than those with less frequency of extension contacts. Henceforth, frequency of extension contacts by development agents with farmers is assumed to be the potential force which accelerates the effective dissemination of adequate agricultural information to the farmers, thereby enhancing farmers' decision to adopt high yielding wheat varieties. From the analysis of marginal effects, increases in one number more frequency of extension contacts by development agents during production season, increased the likelihood of adopting high yielding wheat varieties by 0.09 % adopters' of smallholder farmers. Farmers who have more frequency of extension contacts with development agents update themselves on the availability and arrival of high yielding wheat varieties and aware of its application techniques than those less visited by development workers. The studies conducted by Isaiah *et al.* (2007), Hailu (2008), Solomon *et al.* (2011) and Leake and Adam (2014) found frequency of contacts with extension agent affect positively and significantly adoption decision of smallholder farmers. Similarly, Assefa and Gezahegn (2010), Moti *et al.* (2013) and Sisay (2016) also found that the frequency of extension contact has a significant positive effect on adoption of agricultural technologies.

The availability of timely information on high yielding wheat varieties production was found to have a positive and significant influence on the decision possibility of households' adoption of high yielding wheat varieties. The role of information in decision-making process is to reduce risks and uncertainties, to enable farm households to make right decision on adoption of high yielding wheat varieties. It could be taken as an important source for adopting high yielding wheat varieties in such a way that farmers could easily afford seed and other costs; and farmers are mostly exposed to new and updated information since they access relevant information available in the area. For this this reason, having more access to agricultural information, *citrus paribus*, increases likely use of high yielding wheat varieties by 36.6 % the likelihood of adoption decision of high yielding wheat varieties. This result go along with the study done by Mahdi Egge (2005), Negera and Getachew (2014).

The model output indicated that participating in different off-farm activities having more income earned (the natural log of income earned from non-farm activities) was found to have a positive and significant influence on adoption decision of high yielding wheat varieties. This is believed to raise their financial position to acquire new agricultural inputs. As expected having alternative sources of income from non/off-farm activities, had significant relationship with probability of adoption of high yielding wheat varieties. In the study area, trading *khat*, livestock, and petty trade are the major sources of off-farm income activities in which sample households were participating. The estimated marginal effect of the probit regression model shows, for each additional or unit increase (one Ethiopian Birr) in off-farm income, increases the likelihood of adoption of high yielding wheat varieties by 2.5 % than those did not have off-farm income activities. This study confirms the findings of Habtemariam (2004), Teshale (2006) and Berihun *et al*. (2014).

No matter how well the new technologies work on research stations, if farmers do not have them for use, their development would be in vain. Perception of farmers toward attributes of high yielding wheat varieties is one of the factors that could speed up the change process and adoption of new crop varieties. As being major crop in the study area, high yielding wheat varieties play a vital role in fulfilling household consumption and cash requirements. The result of probit model was showed that perception of farmers toward attributes of high yielding wheat varieties (high yield per unit area, better price of wheat grain, short maturity period, resistance to diseases and lodging) positively and significant affected adoption of high yielding wheat varieties. From the analysis of marginal effect perceiving that high yielding wheat varieties have better yield per unit of area, likely increases the likelihood of adopting high yielding wheat varieties by 45.5 % of adopters' farmers. The results suggest that farmers in the study area seek specific varietal attributes, such as yield potential, tolerance to disease and lodging, better wheat grain price and short maturity period. The finding of farmer perceptions of high yielding wheat varieties-specific characteristics significantly determine adoption decisions and is consistent with evidences in literature, which suggests the need to go beyond the commonly considered socio-economic, demographic and institutional factors in adoption modeling by Feder *et al*.(1985) and Feder and Umali (1993). Similar to this, adoption studies by Wubeneh (2003) and Bayissa (2010) considering farmers' perception of technology attributes have found that attributes determine the adoption choices of farmers. In addition, studies by Adesina and Zinnah (1993) revealed that farmers have subjective preferences for technology characteristics and this could play major roles in adoption.

Farmers' memberships to local organizations, which have an indirect influence on the adoption behavior of farmers. It links the individual to the larger society and exposes households to a variety of ideas. This exposure makes farmers positively predisposed towards innovative ideas and practices. Memberships to organizations have negative and significant influence on adoption of high yielding varieties, implying that farmers who are members of different local organizations are less likely to adopt high yielding wheat varieties. A possible reason for this result relates to the fact that organizations found in the study area are not actively participate in the high yielding wheat cultivation supporting activities and unable to provide credit services for member farmers to purchase high yielding wheat seed and fertilizers since majority are like *Equib, Idir* which serves as a social security support. The primary cooperatives available also facilitate mostly basic supplies as sugar, and edible oil at reasonable price rather than supplying production supporting agricultural inputs such as plant protection chemicals for farmers and purchasing their outputs at better price. The marginal effects result show that being member of local organizations, *citrus paribus*, reduces by 27.8% the likelihood of adoption of high yielding wheat varieties. Contrary to this, Tewodaj *et al.* (2009) and Degnet and Mekibib (2013) found a positive result of relationship of cooperative membership with technology adoption.

5. CONCLUSIONS

The use of improved agricultural technologies particularly high yielding wheat varieties is considered to be the most important input for the achievement of increased agricultural productivity of smallholders in Ethiopia. Wheat contribution to households' food sources (nutritional value), income, feed for livestock, hatching for roofing, and provides job opportunities for farming households and urban dwellers who are engaged in its trading activities.

The study identified key factors that influence adoption process in the study area. This insight is also useful to rethink about the barriers of adoption of high yielding wheat varieties. The prominent variables were categorized as household personal and demographic, socio-economic, institutional and psychological factors. Results confirmed that the likely of adoption decision of high yielding wheat varieties were strongly related to a range of factors, which need due consideration.

Comparisons of the main features of farmers from descriptive statistics results revealed that adopter sample households relatively more educated than non-adopters; have more frequency of extension contacts, owning more land holding, tropical livestock unit, and slightly less family size in the study area. In addition, adopters have better accessibility to information, and good perceptions towards certain attributes of high yielding wheat varieties (high yield per unit area, better market price, short maturity, resistance to diseases and lodging) as compared to non-adopters groups.

The probit model results showed that the contributing factors on the probability of adoption of high yielding wheat varieties were sex of households head, land holding size, tropical livestock unit, frequency of extension contacts, access to agricultural information, off-farm income, perception of farmers' towards attributes of high yielding wheat varieties, and affiliations to local organizations that found to be the most pertinent in in the study area.

Based on results of descriptive statistics and the econometrics models, recommendations are suggested for future research, policy and development intervention activities to promote adoption of high yielding wheat varieties. Therefore, the following recommendations were generalized based on results of this study:

Farmers with more land are more likely to adopt and allocate a relatively higher share of their land for high yielding wheat varieties. Thus, adoption becomes more difficult in the farms with relatively small land size. However, increasing the size of landholding cannot be an option to increase high yielding wheat varieties adoption and market supply since land is a finite resource. Therefore, intervention aimed to improve land fertility status and increasing productivity of land through proper utilization of available land resource is required.

Enhancing the livestock assets of the household as it provides manures for their farm, means of transportation of their produce to the market and provide financial liquidity for the farmers at cash shortages necessary. Therefore, it is recommended that smallholder farmers need to be provided technical support especially medication services of animals since the area is exposed to high prevalence of tsetse fly diseases and improving productivity of livestock for draught power. Therefore, research centers, agricultural development offices and extension workers need to play more role on the livestock production, management, and diseases control through improving resistance of livestock and their nutritional supply.

Promoting and facilitating access to information is found to be one of the key area of intervention. Hence, it is recommended that the district agricultural development office and development endeavors should take measures on the access of overall agricultural information to the farmers mainly timely and relevant information is necessary.

For the effectiveness of the agricultural extension services, an appropriate and effective frequency of extension contacts can encourage farmers to use high yielding wheat varieties. Thus, the researcher suggests that the development agents increase the frequency of extension contacts by identifying farmers' situation and problems that encourages the adoption of high yielding wheat varieties of smallholders.

Off-farm income activities are important through which rural households get additional income and exposure to informal ways of acquiring information. Thus, it is recommended that encouraging households' participation on off-farming activities by creating favorable conditions and better opportunities for smallholders during their off-seasons to have some non-farm income source in addition to their regular farm activities.

Farmers have their own preference criteria for adoption among the available high yielding wheat varieties. The finding of this study suggested that farmers in the area seek specific varietal attributes, such as yield potential, tolerance to disease and lodging, better wheat grain price and short maturity period. Therefore, the research centers and extension system has to give more attention to participatory research which considers farmers' priorities and needs.

6. ACKNOWLEDGEMENTS

My deep sense of appreciation and heartfelt thanks place on record to my advisor Dr. Degye Goshu for his keen interest, intellectual depth, encouragement and insightful criticism and valuable comments for the completion of this write up of manuscripts. I am very grateful to Global Development Network (GDN) and highly indebted to known as Building Ethiopia's Research Capacity in Economics and Agribusiness (BERCEA) Project initiators and coordinators, Dr. Mengistu Ketema and Dr. Degye Goshu for awarding me capacity building in the position of economics and agribusiness and for their timely organization of appropriate materials for filling the gap in applied research and financial support they rendered. I would like to thank my employer Ethiopian Agricultural Research Institute (EIAR) for giving me time logistic and other financial support provided me to continue the study.

7. REFERENCES

Adesina, A.A. and Zinnah, M.M. 1993. Technology characteristics, farmers' perception and adoption decisions: A Tobit Model Application in Siera-Leone. *Journal* of *Agricultural Economics,* 13 (1):1-9.

Ahmed Aliye. 2010. Determinants of adoption of improved durum wheat (*Triticum durum*) varieties in the highlands of Bale: The case of Agarfa district, Ethiopia. MSc thesis, Haramaya University, Haramaya, Ethiopia.

Amhara National Regional State Bureau of Rural Development. 2003. Rural households' socioeconomic baseline survey of 56 Woredas in the Amhara region. Bahir Dar, Ethiopia.

Asfaw Negassa, Gunjal, K., Mwangi, W. and Beyene Soboka. 1997. Factors affecting the adoption of Maize production technologies in Ethiopia. *Ethiopian Journal of Agricultural Economics,* 2: 52-69.

Assefa Admassie and Gezahegn Ayele. 2010. Adoption of improved technology in Ethiopia. Ethiopian. *Journal of Economics,* 1 (5): 155-178.

Bayissa Gedefa. 2010. Adoption of Improved Sesame Varieties in Meisso District of West Hararghe Zone, Ethiopia. MSc Thesis, Haramaya University, Haramaya, Ethiopia.

Bekele Shiferaw, Menale Kassie, Moti Jaleta and Chilot Yirga. 2014. Adoption of improved wheat varieties and impacts on household food security in Ethiopia. *Journal of Food Policy, 44: 272–284.*

Berihun Kassa, Bihon Kassa and Kibrom Aregawi. 2014. Adoption and impact of agricultural technologies on farm income: Evidence from southern Tigray, Northern Ethiopia. *International Journal of Food and Agricultural Economics,* 2 (4): 91-106.

CIMMYT (International Maize and Wheat Improvement Center). 1993. The Adoption of Agricultural Technology: *A Guide for Survey Design,* CIMMYT, Mexico.

CSA (Central Statistical Agency). 2007. Summary and Statistical Report of the 2007 population and housing census. Addis Ababa, Ethiopia.

CSA (Central Statistical Agency). 2015a. Agricultural sample survey of 2014/2015 report on area and production for major crops (private peasant holdings, Meher season). Statistical bulletin No.578. May, 2015. Addis Ababa, Ethiopia.

CSA (Central Statistical Agency). 2015b. Key findings of the 2014/2015 agricultural sample surveys. S.G 1. September, 2015. Addis Ababa, Ethiopia.

Degnet Abebaw. 1999. Determinants of Adoption of HYVs of Maize in Jimma Zone: The case of Mana and Kersa Woredas. MSc Thesis Haramaya University, Haramaya, Ethiopia.

Degye Goshu. 2013. Agricultural technology adoption, diversification, and commercialization for enhancing food security in eastern and central Ethiopia. Doctoral Dissertation, Haramaya University, Haramaya, Ethiopia.

Dorosh, P. and Rashid, S. 2013. *Food and Agriculture in Ethiopia: Progress and Policy Challenges.* University of Pennsylvania Press. Philadelphia, USA.

Feder, G., Just, R.E. and Zilberman, D. 1985. Adoption of agricultural innovation in developing Countries. *Economic Development and Cultural Change,* 33: 255-297.

FAO (Food and Agriculture Organization of the United Nations). 2009. How to feed the world in 2050. (www.fao.org/fileadmin/templates/wsfs/docs/expertpaper/.

Feder, G. and Umali, D. 1993. The adoption of agricultural innovations: A review. *Journal of Economic Development and Cultural Change*, 43: 215-239.

Greene, W. H. and Zhang, C. 2003. *Econometric analysis*. Prentice Hall Upper Saddle River, NJ.

Guerrem, E. and Moon, H. R. 2004. *A study of semi-parametric binary choice model with integrated covariates*. LSTA, Universite Paris 6 and CREST.

Gujarati, N. W. 1995. *Basic Econometrics, 3rd Edition*. McGraw Hill, Inc. New York.

Isaiah, K.O., Fred, U.N. and Washington, W.O. 2007. Socio-economic determinants of adoption of improved sorghum varieties and technologies among smallholder Farmers in Western Kenya.

Habtemariam Abate. 2004. The Comparative influence of intervening variables in the adoption behavior of Maize and Dairy farmers in Shashemene and Debre Zeit, Ethiopia. Doctoral Dissertation, University of Pretoria, Pretoria, South Africa.

Hailu Beyene. 2008. Adoption of improved teff and wheat production technologies in crop livestock mixed system in Northern and Western Zones of Ethiopia. Doctoral Dissertation, University of Pretoria, South Africa.

Hassen Beshir, Bezabeh Emana, Belay Kassa and Jema Haji. 2012. Determinants of chemical fertilizer technology adoption in north eastern highlands of Ethiopia: The double hurdle approach. *Journal of Research Economics and International Finance*, 1 (2): 39-49.

Jon, P. 2007. The adoption and productivity of modern agricultural technologies in the Ethiopian highlands: a cross-sectional analysis of maize production in the west Gojam zone, *proceeding of the 2nd International Symposium on Economic Theory, Policy and Applications*, 6-7 August 2007. Athens, Greece.

Leake Gebresilassie and Adam Bekele. 2015. Factors determining allocation of land for improved wheat variety by smallholder farmers of northern Ethiopia. *Journal of Development and Agricultural Economics*, 7 (3): 105-112.

Mahdi Egge. 2005. Farmers' evaluation, adoption and sustainable use of improved sorghum varieties in Jijiga Woreda, Ethiopia. MSc Thesis, Haramaya University, Haramaya, Ethiopia.

Mahdi Egge, Tongdeelert, P., Rangsipaht, S. and Tudsri, S. 2012. Factors affecting the adoption of improved sorghum varieties in Awbare district of Somali Regional State, Ethiopia. *Journal of Social Science*, 33: 152-160.

Menale Kassie, Bekele Shiferaw and Muricho, G. 2010. Adoption and Impact of improved groundnut varieties on rural poverty. Evidence from rural Uganda. *Environment for Development*, Discussion Paper Series EfD DP 10-11.

Menale Kassie, Bekele Shiferaw and Muricho, G. 2011. Agricultural technology, crop income, and poverty alleviation in Uganda. *World Development*, 39: 1784-1795.

Menale Kassie, Bekele Shiferaw, Mmbando, F. and Muricho, G. 2012. Plot and household level determinants of sustainable agricultural practices in rural Tanzania. *Environment for Development*, Discussion Paper Series EfD DP 12-20.

Moti Jaleta, Chilot Yirga, Menale Kassie, Groote, H.D. and Bekele Shiferaw. 2013. Knowledge, adoption and use intensity of improved maize technologies in Ethiopia. *Paper presented at the 4th International Conference of the African Association of Agricultural Economists*, 22-25 September 2013. Hammamet, Tunisia.

Negera Eba and Getachew Bashargo. 2014. Factors affecting adoption of chemical fertilizer by smallholder farmers in Guto Gida District of Oromia Regional State, Ethiopia. *Journal of Science, Technology and Arts Research*, 3 (2): 237-244.

Nega Gebresilassie and Senders, H. 2006. Farm-level adoption of sorghum technologies in Tigray, Ethiopia. *Journal of Agricultural Systems*, 91 (1-2): 122-134.

Ouma, J., Murithi, F., Mwangi, W., Verkuijl, H., Gethi, M. and De Groote, H. 2002. Adoption of Maize Seed and Fertilizer Technologies in Embu District. CIMMYT, D F, Mexico.

Pindyck, R.S. and Rubinfeld, D.C. 1981. *Econometric models and Econometric factors, 2nd edition.* McGraw/Hill book Co. New York.

Sall, S., Norman, D. and Featherstone, A. M. 2000. Quantitative assessment of improved rice varieties adoption in Senegal: farmers' perspective. *Journal of Agricultural systems,* 66 (2): 129-144.

Shiferaw Feleke and Tesfaye Zegeye. 2006. Adoption of improved maize varieties in southern Ethiopia: Factors and Strategy Options. *Journal of Agricultural Economics*, 31 (5): 442-457.

Sisay Debebe. 2016. Agricultural technology adoption, crop diversification and efficiency of maize-dominated smallholder farming system in Jimma Zone, Southwestern Ethiopia. Doctoral Dissertation, Haramaya University, Haramaya, Ethiopia.

Solomon Asfaw, Bekele Shiferaw, Simtowe, F. and Mekbib Gebretsadik. 2011. Agricultural technology adoption, seed access constraints and commercialization in Ethiopia. *Journal of Development and Agricultural Economics*, 3 (9): 436-447.

Solomon Asfaw, Menale Kassie, Simtowe, F. and Lipper, L. 2012. Poverty reduction effects of agricultural technology adoption: A micro-evidence from rural Tanzania. *Journal of Development Studies*, 48 (9):1288-1305.

Tewodaj, M., Cohen, M. J., Birner, R., Lemma, M., Randriamamonjy, J., Tadesse Fanaye and Paulos Zelekawork. 2009. *Agricultural extension in Ethiopia through a gender and governance lens. Ethiopia Strategy Support Program 2 (ESSP2).* Discussion Paper No. ESSP2 007. IFPRI: Washington D.C.

Tsegaye Mulugeta and Bekele Hundie. 2012. Impacts of adoption of improved wheat technologies on households' food consumption in Southeastern Ethiopia. *Selected Poster prepared for presentation at the International Association of Agricultural Economists (IAAE) Triennial Conference*, 18-24 August 2012. Foz do Iguaçu, Brazil.

Tolesa Alemu. 2014. Socio-economic and Institutional Factors Limiting Adoption of Wheat Row Planting in Ethiopia. Ethiopian Institute of Agricultural Research, Kulumsa Agricultural Research Center. Asella, Ethiopia.

Wubeneh Nega. 2003. Farm-Level Adoption of New Sorghum Technologies in Tigray Region, Ethiopia. MSc Thesis, Purdue University, USA.

Yishak Gecho and Punjabi, N. K. 2011. Determinants of adoption of improved maize technology in Damot Woreda, Wolaita Zone, Ethiopia. *Journal of Extension and Education*, 19: 1-9.

YOUR KNOWLEDGE HAS VALUE

- We will publish your bachelor's and master's thesis, essays and papers

- Your own eBook and book - sold worldwide in all relevant shops

- Earn money with each sale

Upload your text at www.GRIN.com and publish for free